Judging Dairy Cows

by G.C. Humphrey

with an introduction by Jackson Chambers

This work contains material that was originally published in 1916.

This publication is within the Public Domain.

This edition is reprinted for educational purposes and in accordance with all applicable Federal Laws.

Introduction Copyright 2018 by Jackson Chambers

Self Reliance Books

Get more historic titles on animal and stock breeding, gardening and old fashioned skills by visiting us at:

http://selfreliancebooks.blogspot.com/

Introduction

I am pleased to present another title in the "Cattle" series.

The work is in the Public Domain and is re-printed here in accordance with Federal Laws.

As with all reprinted books of this age that are intended to perfectly reproduce the original edition, considerable pains and effort had to be undertaken to correct fading and sometimes outright damage to existing proofs of this title. At times, this task is quite monumental, requiring an almost total "rebuilding" of some pages from digital proofs of multiple copies. Despite this, imperfections still sometimes exist in the final proof and may detract from the visual appearance of the text.

I hope you enjoy reading this book as much as I enjoyed making it available to readers again.

Jackson Chambers

DIGEST

The most successful dairyman is a good judge of dairy cows. Expensive feed, high cost of labor, and rising land values, are making it necessary to improve the herd if dairying is to be profitable. Pages 3-7

The dairy cow is a wonderful milk making machine. Besides the power to produce offspring of her type and breeding, the dairy cow needs good size for her breed, capacity for feed, good dairy disposition, good constitution, a well developed udder, good health, dairy breeding, and large capacity for milk and butter fat production. Pages 7-9

Capacity and temperament are indicated by the triple wedges or triangles which are found on the best dairy cows. One wedge is on the back, one on the side, and the other in front of the shoulders. The base of these triangles indicates capacity, the sharp edge, temperament.
Pages 10-15

No dairy cow is perfect in all parts. In judging a cow deficiencies in digestive capacity, temperament, milk secretion, and constitution, should be considered first. Deficiences in form, breed characteristics, or qualities which please the eye, are not as important as those affecting production. Pages 16-21

The Babcock tester and the milk scale pass final judgment on the cow. Millions of dollars worth of feed are consumed by cows which do not "earn their board and keep." Pages 22-23

The best judges of dairy cows are those who own or manage a good dairy herd and who have made a careful study of judging. A dairyman needs to take every opportunity to take part in judging contests.
Pages 24-25

The score card is a measuring stick by which one practiced in its use can indicate the excellence of the cow. The object of the score card is to train the mind to notice the various parts of the animal on which to base one's judgment. Page 25-28

Community judging contests help to improve dairy cows. They can be easily organized and prove interesting to men and boys alike. Prizes increase the interest in the judging contests. Pages 28-34

Wisconsin is the home of many dairy cows famous for their records of production. A picture of the best producing cow of each of the dairy breeds together with a scale of points for the breed are shown.
Pages 35-42

Judging Dairy Cattle

The successful dairyman of the future, even more than in the past, will need to be a good judge of dairy cattle. The ability to select profitable from unprofitable cows has always been a strong factor in successful dairying; in the future it will be even more important to the success of the men who engage in dairying.

Although the consumption of milk and milk products is constantly increasing and the demand for bred-for-milk-and-

---

A Good Judge of Dairy Cattle

Makes fewer mistakes in buying cattle.
Gets better prices for his surplus stock.
Selects and builds up a herd of cows of uniform size, type, breed and quality.
Receives a higher and more uniform production of milk and butter fat.
Makes greater returns over and above the cost of feed and care.
Uses better sires and secures better calves.
Has better success in feeding and showing cattle at fairs and expositions.
Has greater satisfaction and pleasure in owning and managing a dairy herd.

---

butter-fat-production-cattle growing, the cost of feed and labor and the price of land is also rapidly increasing thus making it all the more necessary for the farmer to have better cows. It is becoming more and more important for him to know the family history of the cow, whether her ancestors were pure bred or grade and whether they were exceptional milkers or just ordinary or even poor producers.

Beef Type

Dairy Type

FIG. 1.—THE DIFFERENCE BETWEEN BEEF AND DAIRY TYPES

The beef animal has straight top and bottom lines, while the dairy cow is wedge shaped.

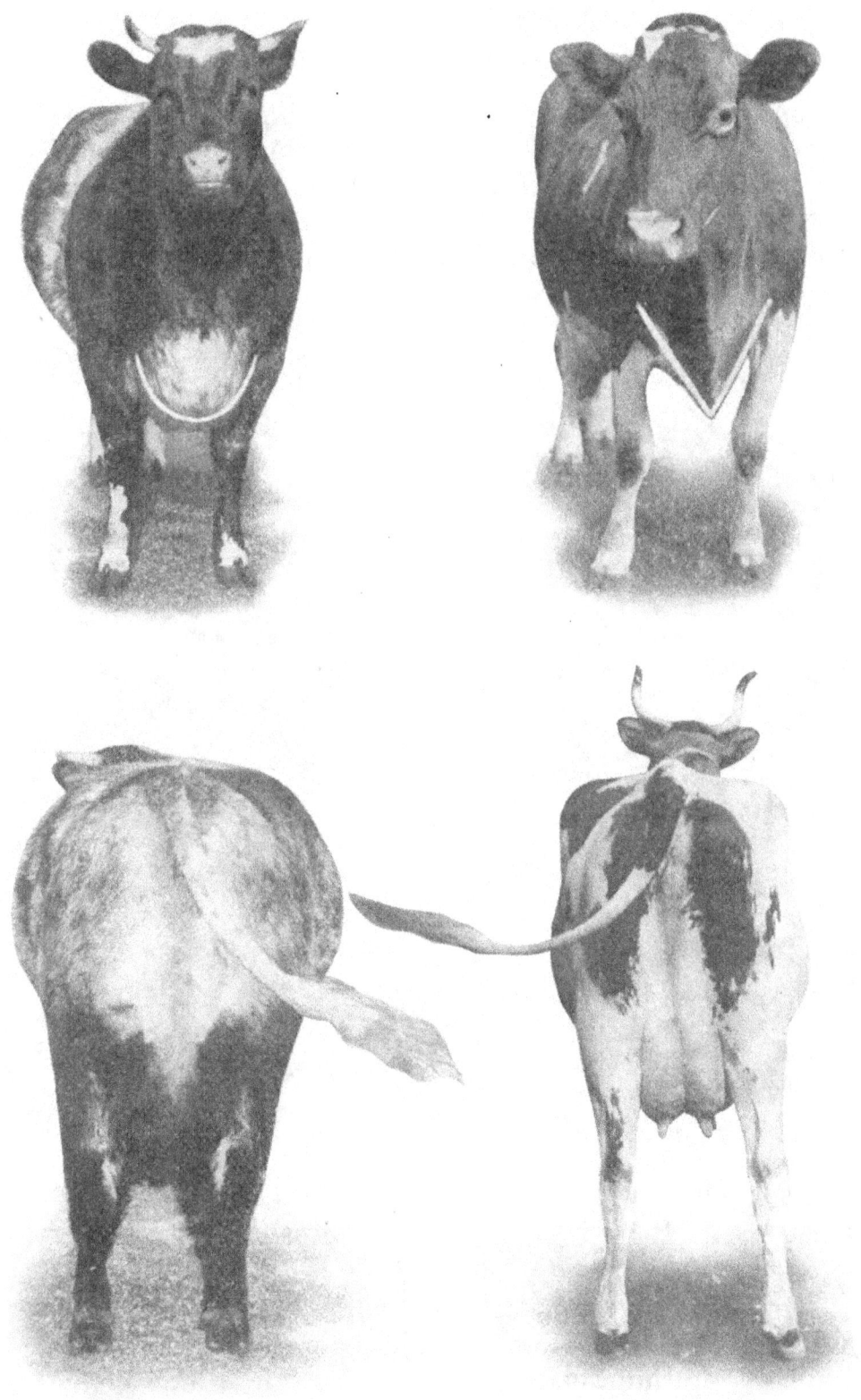

Beef   Dairy

FIG. 2.—BEEF ANIMALS BLOCKY, DAIRY CATTLE ANGULAR

Fullness of the fore and hind quarters are typical of the beef animals. A comparatively long head, sharp brisket, and a pronounced udder development characterize the dairy cow.

WISCONSIN NEEDS BETTER COWS

Wisconsin is destined by virtue of her climate, her location with respect to markets, and her people, to occupy a still more important place in the dairy industry of this country. We need to increase the size of our herds, but we will need even more to improve the production of our

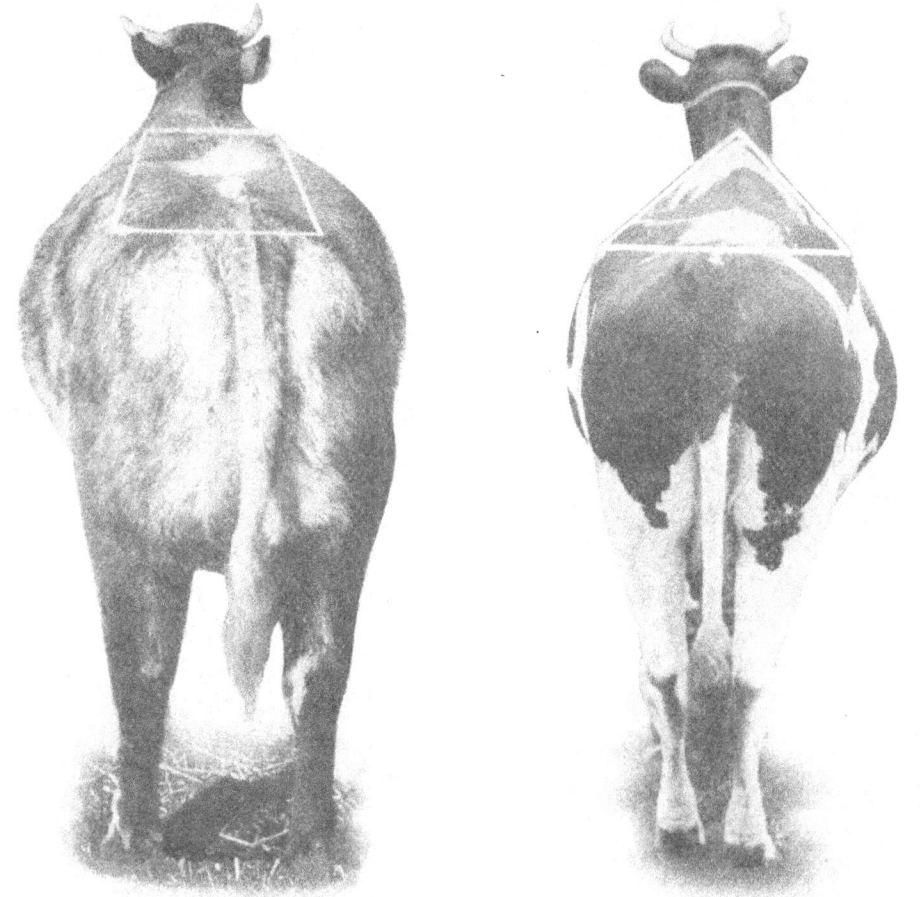

FIG. 3.—BEEF AND DAIRY TOP LINES

Beef animals utilize feed for developing a broad and thickly fleshed back. A triangular shaped back indicates dairy type and milk production.

cows. Great improvement can be made by exercising careful judgment in the selection, breeding, and development of the individuals of a herd.

Cows which are heavy producers usually have certain well defined characteristics, such as conformation, capacity, constitution, and temperament. A ready acquaintance with these, as well as with the line of breeding represented, will aid the dairyman in selecting profit-producing cows. It, of course, should be admitted that even the most expert judges of dairy cattle are unable, by relying solely upon the appearance to the eye and a study of family records, to foretell a cow's ability to produce milk and butter fat. That, in the end, is only told by the use of the milk scales and the Babcock tester.

What is a Good Dairy Cow?

Experiments and experience have shown clearly that as a rule cows which possess certain so-called dairy characteristics are more economical producers of milk and butter fat than those which lack these features. It is then important for the farmer who keeps cows for the milk and butter fat they yield, to see that, as far as possible, his animals have these characteristics.

The dairy cow is best defined by naming her essential features. She may be considered a most wonderful living machine and to be worthy of the name "dairy cow" should have: good size for her breed, good feed capacity, dairy disposition, a good udder, good constitution and health, dairy breeding, large capacity for milk and butter-fat production, and power to produce offspring of her type and breeding.

The cow usually fails in the production of milk to the extent that she fails in one or more of these essential features. Each part of the body shown in Figure 4, bears some relationship to one or more of these essential features and enables one to judge of its prominence and desirability. Where one is able to consider all parts of the body and judge these essential features, he is not likely to err seriously in his judgment.

Cows Need Room for Digestive Organs

A large body, more especially the barrel, in proportion to the size of the animal, indicates capacity. The body of the dairy cow should be wedge shaped as viewed from either the front, the side or the top of the withers. It should be

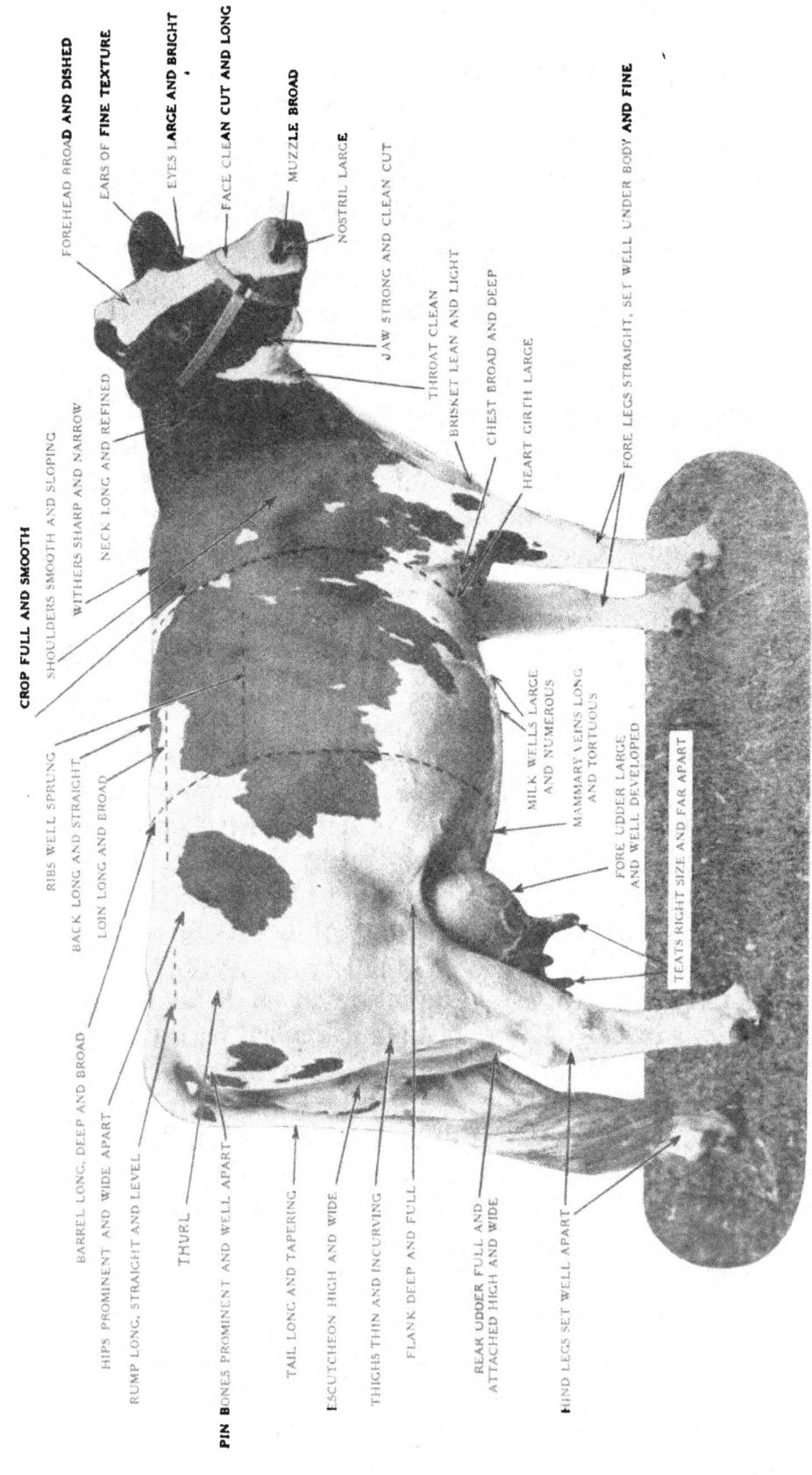

FIG. 4.—THE PARTS OF A DAIRY COW

A dairy cow should have large capacity for feed, a dairy temperament, well developed milk organs, fine quality, perfect health, and be capable of a large production of milk and butter fat. Dutchess Skylark Ormsby, the champion cow for yearly butter fat production, record 27,761.7 pounds milk, 1205.09 pounds butter fat, shows excellence in all parts.

wider at the hip points than at the withers. The floor of the chest between the fore legs should also be wider than the top of the withers. Again, the body should be deeper from the hip points to the bottom of the udder than it is at the fore quarters.

These characteristics of the body have led to the term "triple wedge shape conformation," and in giving consideration to the digestive capacity of the cow, it should be remembered that it is the base ends of the three wedges rather than the sharp ends that indicate feed capacity.

Ribs that are well sprung and far apart, an open chine, a back that is wide over the loins, a large barrel, hips that are wide apart, rear flanks that are full and great depth from hips to lower line of the flank, all combine to indicate large, digestive capacity. A wide forehead, a comparatively long face, broad muzzle, good sized mouth and strong, sinewy jaws, are also considered indications of a large digestive capacity.

The tail is often measured in judging a cow and to meet the standard requirements should reach to, or below, the hocks and carry a good switch. This renders it most useful in brushing flies which is its chief purpose. Excepting as the loose joints of the tail show an open condition of the vertebrae of the back, which is desirable in the dairy cow, it is difficult to understand how the tail would indicate production.

FIG. 5.—LOOK FOR THE WEDGES

The body should be wedge shaped when viewed from the front and top of the withers, wider at the hip bones and at the floor of the chest than at the point of the withers.

Dairy Temperament and Milk Production

The dairy temperament or dairy disposition of a cow indicates her ability to convert feed into milk rather than into flesh. It is a feature which the dairy breeds have acquired through the process of selection and breeding for milk and butter-fat production. It varies in its degree of

FIG. 6.—GREAT DIGESTIVE CAPACITY IS ESSENTIAL

Fullness of flanks and good depth from the hips to the lower line of the rear flank and of the udder, together with well sprung ribs, far apart indicate a large digestive capacity.

strength, even among pure bred dairy animals, and, therefore, needs to be carefully considered in judging. A cow that is a large and econimical producer of milk and butter fat, is almost certain to have a highly developed dairy temperament.

Cows excelling in dairy temperament show the following characteristics:

Features about the head and face are clean cut in outline and indicative of fine quality; eyes are prominent, bright and active; a fine, clean neck, neatly joined to the head, not too full at the throat and comparatively long and thin; shoulders are oblique, comparatively bare of flesh and sharp at the withers; the backbone, hips and pin bones are prominent and sharp; ribs are more or less prominent and open; thighs are thin and incurving, sometimes termed "cat hams;" and bones in all parts of the body indicate quality rather than coarseness.

Sharp Wedges Indicate Temperament

The lean, muscular tissue, on the outside and underneath the shoulder blades and along the back, accounts for the comparatively sharp condition of the withers. The wedge-shaped conformation shown in Fig. 5, is due to the absence of flesh about the neck and the fore quarters. It may be said, therefore, that the sharp end of the triple wedge-shaped conformation is indicative of dairy temperament.

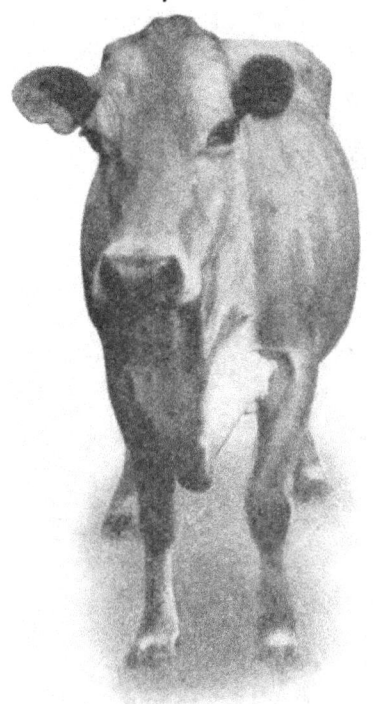

FIG. 7.—DAIRY CHARACTER

A wide forehead and comparatively long face, broad muzzle, and good sized mouth, indicate desirable dairy qualities.

In judging quality and condition of the muscular tissue of the body, consideration should be allowed for the size, age and stage of lactation of the animal. It should also be borne in mind that the bones and muscular tissues in a large cow are naturally heavier than in a smaller or younger animal. Then, too, there is not the natural refinement and spareness of form in the larger breeds that there is in the smaller ones. Marked coarseness, however, in any animal is undesirable and is usually accompanied by a sluggish disposition that in the case of the dairy cow prevents her from "performing at the pail" satisfactorily. Young heifers with their first calves usually carry more flesh than cows of mature form. All cows that are properly fed usually show a higher condition of flesh development toward the close of their lactation and prior to freshening than they do when four or five months advanced in lactation. This should be taken into consideration in judging dairy temperament.

Must Have Well Developed Udder

The udder is the milk secreting organ and its proper development is, therefore, essential. Cows, even of large digestive capacity and of pure dairy breeding, fail to make satisfactory productions when they have poorly developed udders.

The udder consists of two large glands which are more or less distinctly divided to correspond with each of the four

FIG. 8.—A COW WITH MARKED DAIRY TEMPERAMENT

Clean cut features about the head and face, the fine clean neck, the prominence and sharpness of the back bone, hip points and pin bones, the thin, incurving thighs and the clean, fine shanks in this cow are indications of extreme dairy temperament.

teats. The duct of each teat enters a small cavity termed the "milk reservoir." The milk reservoir of each quarter is more or less surrounded by lobes of glands held in position and closely together by connecting tissue. These lobes may be likened to thick bunches of grapes since each lobe has several divisions called lobules, corresponding to the grapes. The lobules are made of small divisions called "alveoli" which correspond to the seeds of grapes. These alveoli consist of small cells surrounded by a fine network of blood vessels and nerves. The milk is secreted by these cells.

The best cows of all breeds have comparatively large udders with equally developed quarters extending well forward underneath the body and a good distance up behind and between the thighs. Swinging or pendulous udders result from poor attachment. Irregularity in the development of the quarters is a criticism to be offered on many udders. The first consideration, however, should be size and quality. The gland tissue should be fine and plastic rather than fatty or coarse and hard.

Good Circulation of Blood Needed

Only when there is a thorough circulation of blood and all parts of the body are active in performing their respec-

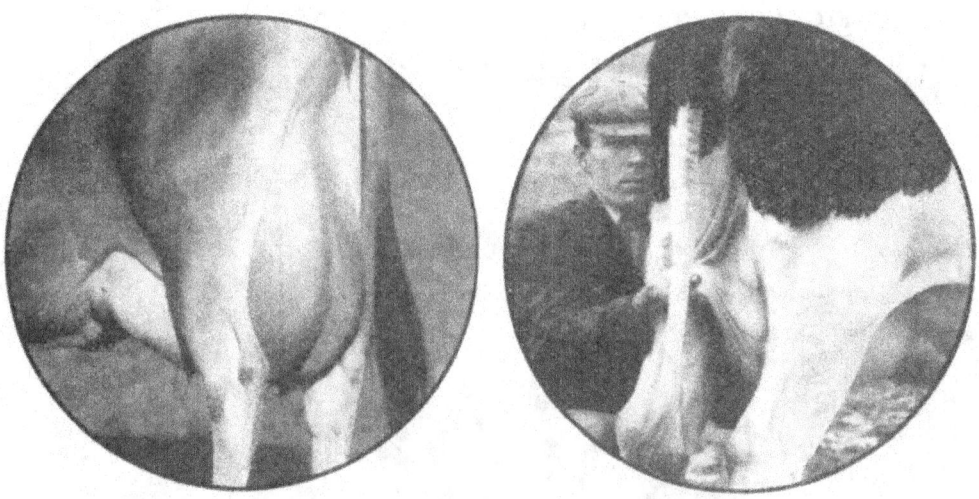

FIG. 9.—TYPES OF GOOD UDDERS

The udders should be large, well proportioned, balanced, extended far forward, and high up between the thighs. (See udder on left). It should be of fine texture, pliable, and the skin should stretch readily when the udder has been milked out. (See udder on right).

tive functions, can the dairy cow be expected to yield a large flow of milk. When the cow is sick, or, by virtue of her poor individuality, is dull and sluggish, there is an inactivity of all the glands of the body, resulting in a dry, harsh condition of the skin, a staring coat and a low production of milk. The circulatory system includes the heart, lungs, arteries and veins. These organs, respectively, force, purify and carry blood to and from all parts of the body.

When the feed which the cow eats is digested and assimilated, the blood carries it to all parts of the body including

FIG. 10.—WELL DEVELOPED MILK VEINS

Large and crooked mammary veins extending far forward on the under side of the body, indicate that the udder is abundantly supplied with blood and capable of large milk production.

FIG. 11.—THE LOCATION OF THE MILK WELLS

Several milk wells of good size through which the mammary veins pass into the body are the best indications of the amount of blood that circulates through the udder and supplies the milk secreting glands.

the glands of the udder which are abundantly supplied with blood vessels. A large amount of blood circulating to the udder is important to milk secretion. The size of the mammary veins and the openings or "milk wells" at the ends of the veins on the underside of the body are the best indications of how much blood passes through the udder. These veins, often called "milk veins," do not carry milk, as some believe, but rather carry blood away from the udder. Blood sometimes becomes gorged in the veins and as a result of too small milk wells, the size of the veins is misjudged. The blood passes into the udder through arteries located deep on the inner sides of the thighs.

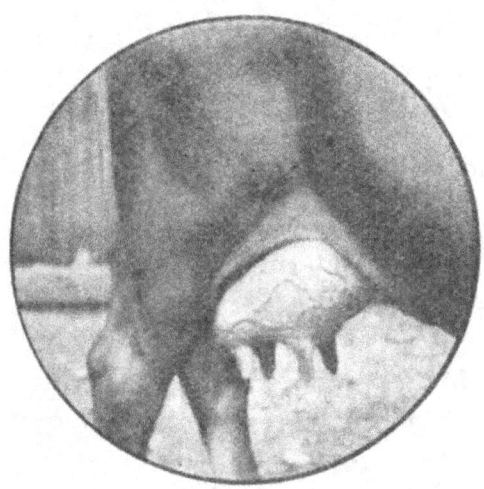

FIG. 12—PROMINENT UDDER VEINS

A good circulation of blood through the udder is indicated by the prominent udder veins. Teats of good size and well placed make hand and machine milking easier.

The oily condition of the skin and the oily secretion noted in the ears and at the end of the tail, are indications of a healthy circulation of blood to all parts of body and a general activity on the part of all healthy glands of the body, including those of the udder. The large, open nostrils providing ample air passages to the lungs for purification of the blood, are also important.

The escutcheon, which is outlined by a mark made by the difference in direction in which the hair runs at the rear of the thighs above the udder, was thought by a French student of the dairy cow, Guenon, to be associated with the artery that carries blood to the udder, and, therefore, indicative of the dairy quality of the cow. If this is true, it should be given as much importance as the milk veins. Guenon also regarded the peculiar condition of spots of hair noted at the back side of the udders of some cows and termed "the thigh ovals," as an important point to consider in connection with the escutcheon. A lack of positive knowledge, however, concerning the relation of these features to milk production does not warrant giving them as much consideration as is given to the milk veins. An escutcheon which is wide and extends high up on the quarters, is considered most desirable and usually is allowed one or two points on the score card for dairy breeds.

No Cow is Perfect in All Parts

No cow ever existed that could be called perfect in all respects when scored by a critical judge. It is expected that every animal will be more or less deficient and the eye of the judge should be quick to note the deficiency. In buying or selling, cows having deficiencies which tend to interfere seriously with their being economical producers should be discarded. Deficiencies in digestive capacity, dairy temperament, milk secretion and constitution should be given first consideration. Deficiencies in symmetry of form, breed characteristics and qualities which simply please the eye are more pardonable than those affecting production. In the show yard, ability to observe and give due consideration to minor deficiencies as well as the more important features, make the work of the judge satisfactory.

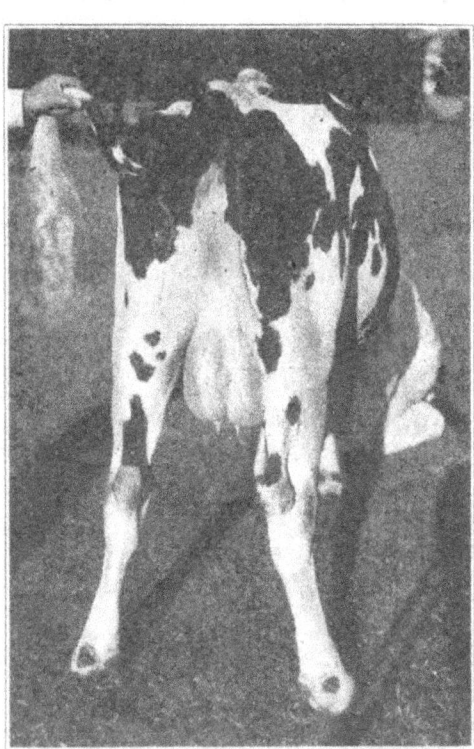

FIG. 13.—ESCUTCHEON AND THIGH OVALS

The escutcheon is outlined by a line formed by the difference in the direction of which the hair lies above the udder. The thigh ovals when found on the rear of each hind quarter of the udder are regarded as indications of a large milk flow.

Some Common Deficiencies

A deficiency in one part of the body is usually accompanied by deficiencies in other parts which one acquiring the art of judging should soon learn. An expert is able to judge the character of an animal quite accurately by taking careful note of the head. A long, narrow head, for example, is usually accompanied by a long, narrow body. Good length of body is desirable in a dairy cow, but a narrow body detracts from digestive capacity. A narrow head usually has small eyes and nostrils and a small mouth.

A small, dull, listless eye expresses inability to do satisfactory work. Small nostrils indicate contracted lung capacity and poor constitution. A small mouth usually goes with small digestive capacity. Marked coarseness of bone, hide and hair are indicative of low productive capacity. Heavy,

FIG. 14.—A SHALLOW BODY LACKS CAPACITY

A narrow head, small eyes, nostrils, and mouth, usually accompany a narrow shallow body. A cow with these characteristics proves a disappointment as a milk producer.

coarse bones over the tops of the shoulders, at the hip points, pin bones, in the tail and legs, are marks of poor dairy temperament.

A body which is short and lacking in depth due to close, short or straight ribs, is objectionable because it detracts from the capacity for feed. The legs of an animal often appear long on account of a small body. The floor of the chest of a cow should be down to a point half way between the knee and elbow joints of the fore legs.

Deficiencies which are common to the hind quarters of the cow, include shortness and narrowness of rump, a drooping rump, narrowness between the thurls and

Weak Back

Roached Back

FIG. 15.—INFERIOR TOP LINES

A straight, strong back is most desirable. Backs which are not straight detract from the general appearance of the cow and may indicate weakness.

pin bones, and thickly fleshed thighs. Narrowness in the hind quarters, especially at the thurls and pin bones, is accompanied by thighs and hind legs which are too close together to permit proper development of the udder. A short rump and thick, heavy thighs are objectionable for the same reason.

There is practically always opportunity for criticising an udder on irregularity of quarters, handling quality, or attachment. An udder does not necessarily have to score perfectly to be capable of making a large production of

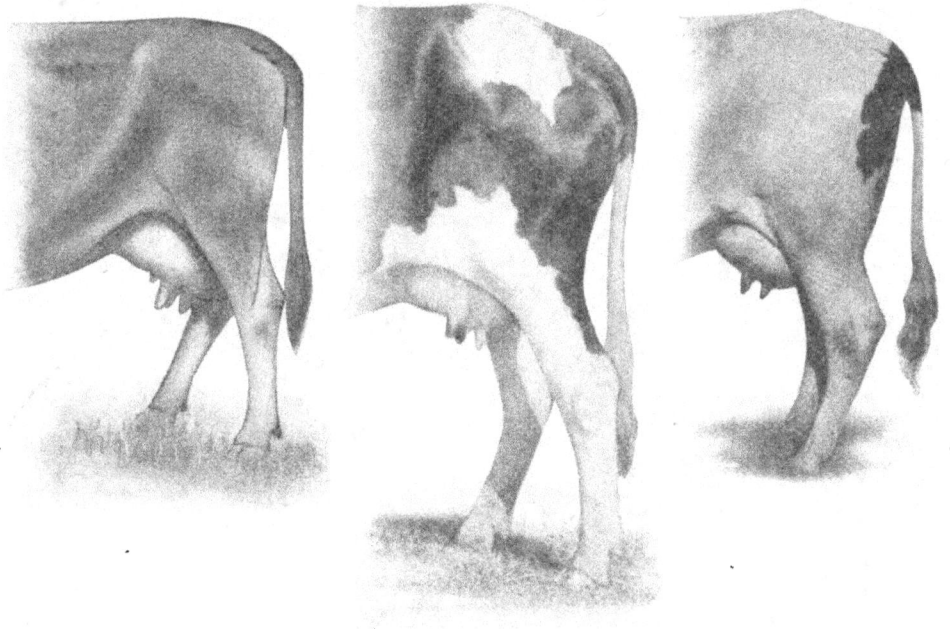

FIG. 16.—THREE TYPES OF BAD RUMPS

Rumps that droop and are low at the pin bones detract from the beauty of the cow and are usually accompanied by udders which tilt forward.

milk, but it is desirable to have it large in proportion to the size of the cow and extend high up behind and well forward in front, with the quarters equally developed and of pliable handling quality. Teats which are too short, too close together or irregularly placed and inconvenient for milking, are often noted.

It is impossible to define perfection in the mammary veins, owing to the great variation in their development. Small, straight veins extending only a short distance forward from the udder and having very few, if any, branches, are characteristic of the veins on poor cows. A network of fair sized veins entering two or more wells on each side of the body, may be considered equal to larger and more prominent veins without branches and extensions. The question is often raised as to why these veins should be crooked. Perhaps no more satisfactory answer can be given than for

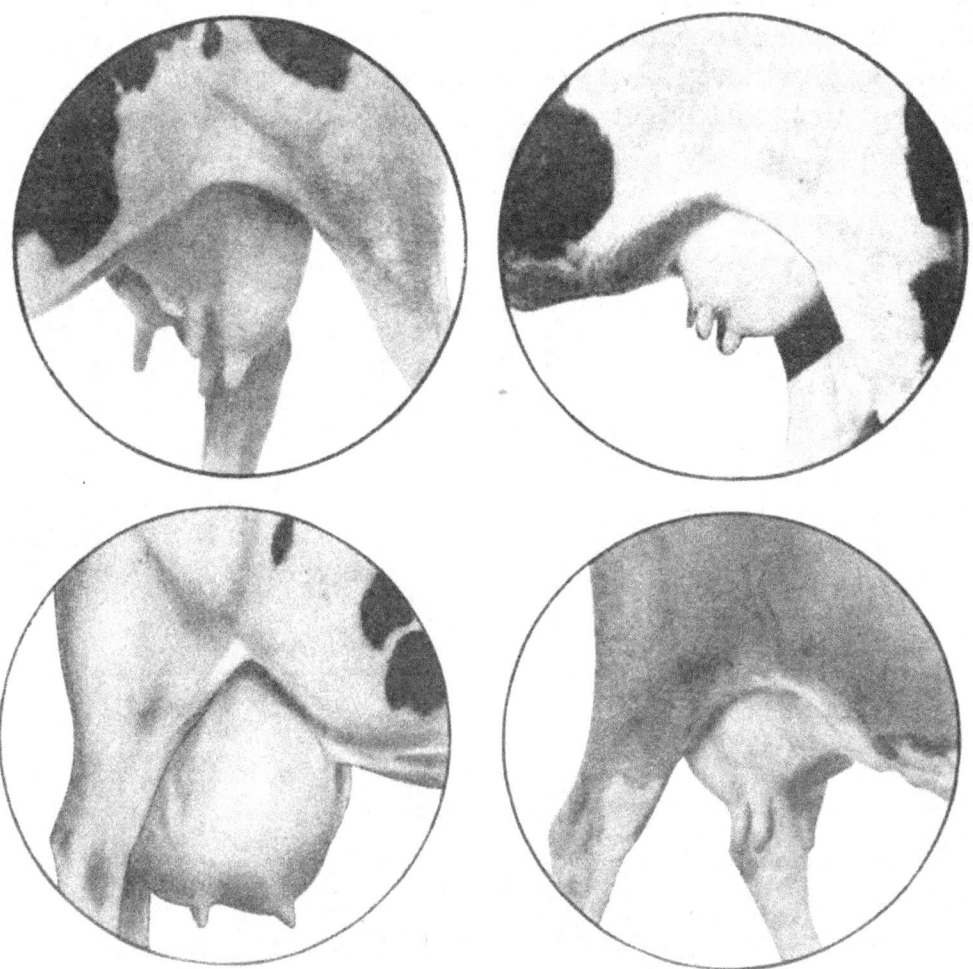

FIG. 17.—FOUR TYPES OF UNDESIRABLE UDDERS

Udders deficient in the fore quarters, irregular in the size of quarters, pendulous in form or funnel shaped make milking hard and reduces the capacity for milk production.

the reason that such veins are commonly found on the best cows.

It is further conjectured that a crooked network of veins indicates the most efficient system of small arteries and veins about the little cells of the udder where the blood gives up the elements which make the milk. The mammary veins should be examined carefully for the reason that in cows considerably advanced in lactation and in young heifers, the veins are never so prominent as in the cow which is in her prime and at the high yielding stage of her lactation. To be able to take into consideration all of the deficiencies of the cow and balance them up against her good qualities, and thus arrive at her real worth for dairy purposes, constitutes the true art of judging the dairy cow.

TYPE, CONSTITUTION AND BREEDING ESSENTIAL

Type which refers to the outline and character of the conformation of an animal indicates or at least suggests its usefulness. In the case of cows it indicates whether they will be useful for the production of milk or the production of beef, or whether they are of any use for the production of either beef or milk. Dairy type refers to an animal having the essential features of the dairy cow. Having these features well fixed in mind makes it possible to judge quickly the desirability of cows for dairy purposes.

A strong constitution is highly essential for the reason that the work of the cow is strenuous when she is fed for maximum production. A cow lacking vitality is rarely ever a good feeder, and consequently is unsatisfactory for milk production. The cow which has a constitution to enable her to be useful for a period of 10 or 12 years in making a large production of milk and regularly bearing offspring, is most profitable.

The breeding or ancestry of a cow largely determines her characteristics, the use she makes of her feed and the characteristics of her calves. The dairy cow by virtue of her breeding, has the characteristics of some one of the recognized dairy breeds. Dairy breeding insures against disappointment when one buys or raises a cow for milk production.

The National Dairy Show Association of this country recognizes Holsteins, Guernseys, Jerseys Ayrshires, and Brown Swiss as the leading dairy breeds. Pure bred animals possess 100 per cent of the blood of their respective breeds. Grade animals have a predominance of the blood of a given breed but less than 100 per cent. Grade cows are usually by pure bred sires and out of native or grade dams.

Must Consider Records of Production

The cow is very much like a race horse when it comes to judging her ability to perform. Both the race horse and the cow must be judged by means of the eye assisted by the record of performance. The milk scale and the Babcock tester assisted by judgment of the eye, are the best means of exercising judgment in building up a profitable dairy herd. Persistently following this means of judging will lead to the establishment of a herd which is both pleasing to the eye and capable of a large and profitable production of milk and butter fat. Every dairyman can afford to weigh the milk from each cow at each milking and have a sample of the milk of each cow tested once a month. Results accurate enough for all practical purposes may thus be secured.

FIG. 18.—AN IMPARTIAL AND UNFAILING JUDGE

Profitable production of milk and butter fat should be the first consideration in breeding and building a dairy herd. This is determined by weighing the milk from each cow at each milking and having a sample of the milk from each cow tested once a month.

Scale and Tester Would Increase Profits

If this means of judging were employed on every dairy farm, feed worth millions of dollars now being eaten by cows which do not pay their cost of keep, would be saved annually or converted into milk and the value of dairy products of the state would be greatly increased.

The Advanced Registry system, maintained by dairy cattle associations, records the milk and butter fat production of cows officially tested, and renders valuable aid in judging pure bred dairy animals on the basis of their ability to perform.

The difference in the production of cows is shown by the annual production and returns of three classes of cows at one time in the University dairy herd.

Class A shows the average annual production and returns per head from the best four cows of the herd for four consecutive years. Class B shows the same data for the poorest four cows kept for a period of four years. Class C shows the average results of four cows which were too poor to keep in the herd for longer than one year during this period.

TABLE I.—DIFFERENCE IN THE PRODUCTION OF COWS

	Class A		Class B		Class C	
	Amt. lbs.	Value	Amt. lbs.	Value	Amt. lbs	Value
Milk	9,984.0		7,478.0		4,929.0	
Skimmilk.........@ 20	7,987.0	$ 15.97	5,982.0	$ 11.96	3,944.0	$ 7.89
Butter fat.........@ 30	426.9	128.07	301.8	90.54	195.8	58.74
Total, a cow a year		$144.04		$102.50		$66.63
Feed cost		73.40		60.32		47.62
Return over feed cost		70.64		42.18		19.01
Cost of purchased feed		23.18		18.30		10.47
Return over purchased feed		120.86		84.20		56.16

Feed prices a ton—hay $20.00; silage $3.50; soilage $3.00; sugar beets $4.00; bran $25.00; oats $30.00; corn $24.00; oil meal $40.00; distiller's grains $30.00; miscellaneous $30.00; pasture, a season $10.00.

The average annual production and returns for four years of the best four cows was $144.04 a cow, of the poorest four cows $102.50, and of the four cows that were too poor to keep more than one year $66.63.

Only by eliminating the poorer cows of the herd can a high herd average be maintained. A breeder of dairy cattle who is particularly anxious to improve the quality and excellence of the herd will be careful to judge dairy cattle by means of the eye assisted by a careful consideration of milk and butter fat production. Furthermore, attention should be given to the pedigree of the animals.

A GOOD FAMILY TENDS TO INSURE GOOD COWS

The pedigree of an animal is a record of its ancestors, or family. The ordinary pedigree usually shows the ancestors for five or six generations. The value of the pedigree lies in the fundamental law of nature that "like produces like." Where the ancestors of a given animal are known to be good, one can judge more accurately than by the eye alone. The careful dairyman, who is anxious to build up the best possible herd, will find it of advantage to study carefully the individuality, pedigree and performance of the cattle.

A Keen Eye Needed for Judging

While it cannot be depended upon alone, the importance of the eye as a means of judging the dairy cow cannot be over-estimated. A well trained eye is a necessity in all stock breeding. It has been well said, "Beauty is bought by judgment of the eye."

Neatness and symmetry of form require that the animal be well balanced and as nearly ideal as possible in all parts. This does not necessarily mean that it will outclass in production and durability animals which have the essential features but are of plainer type. The show yard animal, however, which can combine milk production and pedigree with individuality, is always most pleasing to the eye, and commands the highest price. When its type is more thoroughly established by a greater number of ancestors of like character, and of high productive capacity, there is no reason why it should not reproduce itself and replace the less desirable types.

How to Become an Expert Judge

One may learn a great deal and acquire much of the art of judging from books on the subject and by observing the work of expert judges whenever there is opportunity to do so. The best judges, however, are men who have owned or managed a first-class dairy herd, and who have made a careful study of judging in the show ring. Making careful and accurate observations and exercising judgment based on the best standards of excellence, will assist more than any other thing in becoming an expert judge of live stock. One should take advantage of every opportunity to observe the work of expert judges and to take part in scoring and judging exercises whenever possible.

The Score Card Aids in Learning to Judge

The dairy score card is an enumeration of all parts of the cow, arranged in a given order, with a statement of the requirements and the number of points for a perfect score of each part. The sum of the points for all parts totals 100. This arrangement is termed "a scale of points" and is the standard of excellence by which the individuality of the cow with reference to her form and body characteristics may be judged. Each national breeders' association has a scale of points for its particular breed.

The score card applied to the cow, becomes the measuring stick, so to speak, by means of which one practiced in its use can indicate the excellence of the cow by the total score. A perfect animal would have a total score of 100. There is usually, however, a fault with some part. There may be a general deficiency in many or perhaps in only one or two parts.

It is necessary for the judge to have the score card for the particular class of cows upon which he is to pass judgment thoroughly in mind, if he hopes to do his work sucessfully. Otherwise he will have no standard to guide him and his judgment is very likely to err seriously.

If one were to adhere strictly to the score card method of judging and had two or more cows to judge, he would score each animal separately and rank them in the order of the respective scores. For example, the best animal might score 95, the second best 90, and the third only 75. On account of the great amount of time required to go through the mechanical operation of scoring a large number of animals, the score card system of judging is not practiced at shows or in buying animals.

The primary object of having a score card is to furnish a statement which will systematically train the mind to give consideration to each part of the animal and to formulate some judgment. After practice with the score card, one can pass rapidly over a class of animals and select those which approach nearest the standard. Having selected one which appears nearest to perfection, other animals in the class or herd may be ranked by means of comparison with the first.

In buying cows where one is obliged to take perhaps only one from each of several herds and is anxious to get uniformity of type and quality, it is of great importance and value to have the points of the score card well in mind. It would be well for all dairymen to practice scoring. Score

SCALE OF POINTS FOR DAIRY CATTLE

Student...Date.........................
Animal..Animal................................

GENERAL APPEARANCE—A dairy cow should weigh not less than 800 pounds, have large capacity for feed, a dairy temperament, well developed milk organs, fine quality and perfect health, and be capable of a large production of milk and butter fat.

SCALE OF POINTS	Perfect	Points deficient		Points deficient	
		Student's score	Corrected	Student's score	Corrected
INDICATION OF CAPACITY FOR FEED—25 POINTS					
Face, broad between the eyes and long; muzzle clean cut; mouth large; lips strong; lower jaws lean and sinewy	5				
Body, wedge shape as viewed from front, side and top; ribs, long, far apart and well sprung; breast full and wide; flanks, deep and full	10				
Back, straight; chine, broad and open; loin broad and roomy	5				
Hips and thurls, wide apart and high	5				
INDICATION OF DAIRY TEMPERAMENT—25 POINTS					
Head, clean cut and fine in contour; eyes, prominent, full and bright	3				
Neck, thin, long, neatly joined to head and shoulders and free from throatiness and dewlap	4				
Brisket, lean and light	2				
Shoulders, lean, sloping, nicely laid up to body; points prominent; withers sharp	4				
Back, strong, prominent to tail head and open jointed	3				
Hips, prominent, sharp and level with back	3				
Thighs, thin and incurving	4				
Tail, fine and tapering	1				
Legs, straight; shank fine	1				
INDICATION OF WELL DEVELOPED MILK ORGANS—25 POINTS					
Rump, long, wide and level; pelvis roomy	3				
Thighs, wide apart; twist, high and open	3				
Udder, large, pliable, extending well forward and high up behind; quarters, full, symmetrical, evenly joined and well held up to body	15				
Teats, plumb, good size, symmetrical and well placed	4				
INDICATIONS OF STRONG CIRCULATORY SYSTEM, HEALTH, VIGOR AND MILK FLOW—25 POINTS					
Eyes, bright and placid	2				
Nostrils, large and open	3				
Chest, roomy	5				
Skin, pliable; hair, fine and straight; secretions, abundant in ear, on body and at end of tail	7				
Veins, prominent on face and udder; mammary veins, large, long, crooked and branching; milk wells large and numerous	7				
Escutcheon, wide and extending high up	1				
Total	100				

cards are furnished by dairy cattle breeders' associations and it is a splendid thing for local associations of dairymen to have scoring contests.

Score Card Practice Helps

Every community interested in the improvement of its dairy cattle might have one or more meetings each year for score card practice. Someone with experience and a knowledge of correct dairy form should be chosen for leader and demonstrator.

Here are a few general rules which may be followed: have a sufficient number of animals to avoid having too many men crowd about and score the same one; have the animals stand at ease on a level floor or piece of ground provided with good light; inspect the cow from all directions at a distance of 10 to 16 feet noting carefully her size, form, quality, and alertness, (too close contact with the animal often leads a judge to be deceived); note each point in the order it is named on the score card; use the hand only to determine the quality of the hair and hide, the secretion of the skin, the openness of the back, distance apart of the ribs, the condition of the mammary veins and milk wells, and the quality of the udder.

The following rules may be employed to determine the extent to which any part should be cut. The cut for a deficieny never exceeds half of the total number of points allowed for perfection and is never less than .25 of one point. For example, the face is allowed 5 points for perfection and however deficient is never cut more than 2.5, nor less than .25 if deficient in the least. The judgment of the student must decide the amount of discount between these limits. No cut is made where no deficiency is noted. When all have completed their scores of the animal or group of animals, the leader or expert should read his score and call for comparisons and discussions on all parts of the animal to help all to arrive, as far as possible, at a proper judgment.

The form of score card shown on page 26 is used more especially in teaching elementary stock judging at the University of Wisconsin. It teaches the essential features and the structural requirements of the dairy cow. It applies to all dairy cows without reference to breed and will be valuable to anyone interested in the selection and judging of cows, who is not familiar with what constitutes dairy type and the essential features of the dairy cow.

A scale of points for each dairy breed is also published on pages 35–42. These have been prepared by the respective breeders' associations and apply only to the breed of cattle for which each was prepared. They should be carefully studied by anyone who attempts to select and breed high

FIG. 19.—LEARN TO JUDGE BY JUDGING

The people of many Wisconsin communities have improved their herds by holding dairy cattle demonstrations and judging contests.

grade and pure bred dairy cattle. They teach the size, color markings and pecularities of form of the different breeds as well as the necessary features which a dairy cow must have. The breeder and judge of dairy cattle should thoroughly familiarize himself with the scale of points for the particular breed or breeds he may have occasion to judge.

Hold Judging Exercises and Contests

When one becomes familiar with all the parts and essential features of the dairy cow, competitive judging or placing a group of animals in the order of their merit, will be interesting and helpful in acquiring the art of judging. Judging exercises and contests can be held at meetings of cattle breeders' associations, county and state fairs, farmers' clubs, boys' clubs and various other meetings. Officials of local breed associations and fairs, county agricultural representatives, high school instructors in agriculture, leaders of boys' clubs and anyone else interested in the bettering of rural conditions can organize and hold judging contests.

ORGANIZATION OF JUDGING CONTESTS

Someone must be superintendent and responsible for organizing the contest and having everything accomplished in regular order. Responsible committees should be appointed to assist and make the following arrangements:

A definite time and place for meeting; a sufficient number of contestants to take part in the contest and make it worth while and interesting; several suitable classes of cattle for judging; and a competent judge or committee to pass finally on the respective classes of animals and assist the superintendent in rating the work of the contestants.

The contestants should be grouped as far as possible to avoid unfair competition. For example, small boys up to a certain age limit should be given an opportunity to compete with one another and not with older and more experienced boys.

Individuals and teams from graded schools, high schools and agricultural schools, boys from 16 to 25 years old not in school, men over 25 years old, breeders of grade cattle, breeders of pure bred cattle and even women and girls may constitute interesting groups to take part in judging contests. The contestants should be properly enrolled and, where prizes are to be awarded, given a number to be used in place of their names on report cards and by which they will be designated throughout the contest.

It is best to select the classes of animals to be judged just prior to the beginning of the contest and without any suggestions from any of the contestants or their coaches. On page 32, is shown a scheme for grading the contestants on their placing a class of four animals, and, therefore, choosing this number of animals for each class will greatly assist in the work of rating contestants.

[1] Reasons should be given by the contestants for the order in which they place the first three animals by mentioning the particularly strong or weak features which characterize the animals concerned. It is customary in judging contests to allow contestants for their final score, 60 per cent of their credits on placing and 40 per cent of their credits for reasons.

Premiums and Rewards Add Interest

To offer prizes for which contestants can compete gives added interest to judging contests. The right kind of a committee can usually solicit and secure numerous articles that make good prizes and that may range from a small cash prize or a small article which the village storekeeper is willing to donate, to a good cash prize or a silver loving cup. It is well to have contestants win a loving cup two or more years in succession before claiming it permanently. The primary object of judging contests, however, should be to promote better judgment and the greatest prize will be the development of ability to select cattle that will prove most valuable to their owners.

Suggested Grades for Judging Contestants

To use this system in rating contestants placing classes of four animals in all possible ways, number the animals of each class 1, 2, 3, 4 and stand them in any order desirable. Cutting the numbers out of an old calendar will be an easy way to provide numbers. Take the result of the judging committee when it has finally passed upon a given class, which for example may be 3, 2, 1, 4, and write it vertically opposite A, B, C, D, which represents the correct placing and a grade of 100.

$$
\begin{aligned}
\text{Example—A–3} \\
\text{B–2} \\
\text{C–1} \\
\text{D–4} \\
\hline
\text{Grade— 100}
\end{aligned}
$$

[1] The competitive judging card shown on page 33 can be furnished by the College of Agriculture at minimum price.

Contestants whose work agrees with the judging committee will be graded 100. For contestants whose work does not agree with that of the judging committee, transpose the numbers on their report cards in the order given to letter opposite the corresponding number, indicating correct placing and note the grades for the result secured.

Example:

Correct placing	Incorrect placings	
A–3	A–3	A–3
B–2	C–1	B–2
C–1	B–2	D–4
D–4	D–4	C–1
Grade—100	75	85

The system will be very helpful in quickly and consistently grading the work of contestants on placing a class of animals. The grades are based on discounts of 20 and 15 respectively, for each place the first and second prize animals are placed below their proper ranks; 10 for each place the third prize animal is placed above or below its proper rank, and 5 for each place the fourth prize animal is placed above its proper rank. The grades may appear somewhat inconsistent in a few instances, although they are apt to be much more so in rating the work of contestants if one does not have a definite system to follow. One might consider a discount of 20 on misplacing the first two animals too severe, especially if the difference between the two animals were only slight. The contestant, however, who can discriminate on small differences and arrive at correct judgment, deserves much more credit than he who errs on such differences.

It is impossible to formulate any pre-arranged scheme for grading students on reasons given for placing. This will have to be done in an arbitrary manner by the committee which examines the reports of contestants. Clear statements which point out the discriminating features one animal possesses over another deserve full credit. Rambling expressions which fail to note differences in essential features should be discounted heavily.

Grades On Placing

A	A	A	A	A	A
B	B	C	D	D	C
C	D	B	B	C	D
D	C	D	C	B	B
100	85	75	65	60	55

B	B	B	B	B	B
A	A	C	D	D	C
C	D	A	A	C	D
D	C	D	C	A	A
80	65	50	40	30	25

C	C	C	C	C	C
A	B	A	D	B	D
B	A	D	A	D	B
D	D	B	B	A	A
45	40	25	20*	15	5*

D	D	D	D	D	D
A	A	B	B	C	C
B	C	A	C	A	B
C	B	C	A	B	A
40	35	35	25	5	0

* Exceptions to the general rule.

Statement of Student in Competitive Judging

Student.. Date......................
Judging........................... Class................ Section......................

| First place | Live weight | Remarks...................................... |

| Second place | Live weight | Remarks...................................... |

| Third place | Live weight | Remarks...................................... |

Dairy Cattle Comparison Card

Student..

Date..

	1st place	2nd place	3rd place	4th place	5th place
Dairy temperament					
Capacity					
Skin					
Constitution					
Top line					
Head					
Neck					
Shoulder					
Body					
Rump					
Veining					
Fore—udder					
Rear—udder					
Teats					
Udder					
Breed Characters					
Placing					

FIG. 20.—THE NOTED SCOTCH BREED

Ayrshire cow Kilnford Bell 3rd, No. 30643. This is a Wisconsin cow that was twice grand champion dairy cow at the National Dairy Show. She also has a semi-official yearly record of 13,525 lbs. milk, testing 3.94 per cent, and 532.98 butter fat.

The Ayrshire cow Lily of Willowmoor, holds the world's record for this breed, with a production of 22,596 lbs. of milk testing 4.22 per cent and 955.56 lbs. butter fat.

The native home of the Ayrshire is southwestern Scotland. Cattle of this breed were imported into Canada early in the 19th century and into the United States in 1822. The American Ayrshire Breeders' Association, organized in 1875, provides for the registration of Ayrshire cattle bred in the United States. C. M. Winslow, Brandon, Vt., is the present secretary.

SCALE OF POINTS FOR AYRSHIRE COWS

Head..10

    Forehead—Broad and clearly defined.. 1
    Horns—Wide set on and inclining upward.. 1
    Face—Of medium length, slightly dished; clean cut, showing veins.......... 2
    Muzzle—Broad and strong without coarseness, nostrils large.................. 1
    Jaws—Wide at the base and strong... 1
    Eyes—Full and bright with placid expression... 3
    Ears—Of medium size and fine, carried alert.. 1

Neck.—Fine throughout, throat clean, neatly joined to head and shoulders, of good length, moderately thin, nearly free from loose skin, elegant in bearing.. 3

Fore quarters... 10
    Shoulders—Light, good distance through from point to point but sharp at withers, smoothly blending into body... 2
    Chest—Low, deep and full between and back of fore legs............................. 6
    Brisket—Light.. 1
    Legs and Feet—Legs straight and short, well apart, shanks fine and smooth, joints firm; feet medium size, round, solid, and deep................. 1

Body... 13
    Back—Strong and straight, chine lean, sharp and open-jointed.................. 4
    Loin—Broad, strong and level.. 2
    Ribs—Long, broad, wide apart and well sprung.. 3
    Abdomen—Capacious, deep, firmly held up with strong muscular development.. 3
    Flank—Thin and Arching.. 1

Hind Quarters... 11
    Rump—Wide, level, long from hooks to pin bones, a reasonable pelvic arch allowed.. 3
    Hooks—Wide apart and not projecting above back nor unduly overlaid with fat... 2
    Pin Bones—High, wide apart... 1
    Thighs—Thin, long and wide apart.. 2
    Tail—Fine, long and set on level with back... 1
    Legs and Feet—Legs strong, short, straight, when viewed from behind and set well apart; shanks fine and smooth, joints firm, feet medium size, round, solid and deep.. 2

Udder.—Long, wide, deep but not pendulous, nor fleshy; firmly attached to body, extending well up behind and far forward; quarters even; sole nearly level and not indented between teats, udder veins well developed and plainly visible... 22

Teats.—Evenly placed, distance apart from side to side equal to half the breadth of udder, from back to front equal to one-third the length; length 2½ to 3½ inches, thickness in keeping with length, hanging perpendicular and not tapering... 8

Mammary Veins.—Large, long, tortuous, branching and entering large orifices.. 5

Escutcheon.—Distinctly defined, spreading over thighs and extending well upward... 2

Color.—Red of any shade, brown, or these with white; mahogany and white, or white; each color distinctly defined. (Brindle markings allowed but not desirable)... 2

Covering.. 6
    Skin—Medium thickness, mellow and elastic... 3
    Hair—Soft and fine... 2
    Secretions—Oily, of rich brown or yellow color.. 1

Style.—Alert, vigorous, showing strong character; temperament inclined to nervousness but still docile.. 4

Weight.—At maturity not less than one thousand pounds.. 4
                                                                               100

FIG. 21.—ORIGINALLY FROM SWITZERLAND

Brown Swiss cow Merry Merney, No. 33179. Her production of 15,679 lbs. of milk, testing 4.01 per cent, and 628.8 lbs. of butter fat is the highest record for a Wisconsin cow of her breed.

College Bravura 2d, No. 2577, holds the world's record for this breed with 19,460.6 lbs. of milk, testing 4.10 per cent and 798.16 lbs. of butter fat.

Brown Swiss cattle were first imported into America from Switzerland, their native home, in 1869. The Brown Swiss Cattle Breeders' Association of America, was organized in 1880 and promotes the interest of the breed in this country. Ira Inman, Beloit, Wis., is the present secretary.

SCALE OF POINTS FOR BROWN SWISS COWS AND HEIFERS

Head, medium size, and rather long	2
Face, dished, narrow between horns and wide between eyes	2
Ears, large, fringed inside with light colored hair, skin inside of ear a deep orange color	2
Nose, black, large and square, with mouth surrounded by mealy colored band, tongue black	2
Eyes, moderately large, full and bright	2
Horns, short, regularly set with black tips	2
Neck, straight, throat clean, neatly joined to head, shoulders of good length, moderately thin at the withers	4
Chest, low, deep and full between and back of fore legs	6
Back, level to setting of tail, and broad across the loin	6
Ribs, long and broad, wide apart and well sprung with thin, arching flanks	3
Abdomen, large and deep	5
Hips, wide apart, rump long and broad	4
Thighs, wide, quarters not thin	4
Legs, short and straight with good hoofs	2
Tail, slender, well set on, with good switch	2
Hide of medium thickness, mellow and elastic	3
Color, shades from dark to light brown, at some seasons of the year grey; white splashes near udder not objectionable, light stripe along back. White splashes on body or sides objectionable. Hair between horns usually lighter shade than body	4
Fore udder, wide, deep, well rounded but not pendulous, nor fleshy, extending far forward on the abdomen	12
Rear udder, wide, deep, but not pendulous, nor fleshy, extending well up behind	12
Teats, rather large, set well apart and hanging straight	8
Milk veins large, long, tortuous, elastic and entering good wells	6
Disposition quiet	2
Size, evidence of constitution, and stamina	5
	100

FIG. 22.—DEVELOPED ON ONE OF THE CHANNEL ISLANDS

Guernsey cow Yerksa's Top of Gold's Fannie, No. 22362. This cow has completed her semi-official year's record with 19,794.9 lbs. of milk and 981.53 lbs. of butterfat.

Murne Cowan, No. 19597, holds the world's record for this breed with a production of 24,008 lbs. milk, testing 4.57 per cent, and 1098.18 lbs. of butter fat.

Guernseys have their origin in the Islands of Guernsey and Alderney, off the coast of France in the English Channel. They were imported into America as early as 1818. The American Guernsey Cattle Club was organized in 1877, but not until 1893 did wide public interest in this breed develop. Wm. H. Caldwell, Peterboro, N. H., is the present secretary.

SCALE OF POINTS FOR GUERNSEY COWS

Dairy temperament Constitution 38	Clean cut, lean face; strong sinewy jaw; wide muzzle with wide open nostrils; full, bright eye with quiet and gentle expression; forehead long and broad	5
	Long, thin neck with strong juncture to head; clean throat. Back bone rising well between shoulder blades; large rugged spinal processes, indicating good development of the spinal cord	5
	Pelvis arching and wide; rump long; wide, strong structure of spine at setting of tail. Long thin tail with good switch. Thin incurving thighs	5
	Ribs amply and fully sprung and wide part, giving an open relaxed conformation; thin arching flanks	5
	Abdomen large and deep, with strong muscular and naval development, indicative of capacity and vitality	15
	Hide firm yet loose, with an oily feeling and texture, but not thick	3
Milking marks denoting Quantity of flow 10	Escutcheon wide on thighs; high and broad, with thigh ovals	2
	Milk veins long crooked, branching and prominent, with large or deep wells	8
Udder formation 26	Udder full in front	8
	Udder full and well up behind	8
	Udder of large size and capacity	4
	Teats well apart, squarely placed, and of good and even size	6
Indicating Color of milk 15	Skin deep yellow in ear, on end of bone of tail, at base of horns, on udder, teats and body generally. Hoof, amber colored	15
Milking marks denoting Quality of flow 6	Udder showing plenty of substance but not too meaty	6
Symmetry and size 5	Color of hair a shade of fawn, with white markings. Cream colored nose. Horns amber colored, small, curved and not coarse	3
	Size for the breed.—Mature cows, four years old or over, about 1,050 lbs.	2
		100

FIG. 23.—FIRST BRED TO MEET FRIESLAND'S NEEDS

Holstein-Friesian cow Colantha 4th's Johanna, No. 48577. At one time this cow was the world's champion cow and her production now exceeds that of any cow of her breed in Wisconsin. Her yearly semi-official record was 27,432.5 lbs. of milk, testing 3.64 per cent and 998.56 lbs. butter fat. The present world's champion cow, is Duchess Skylark Ormsby No. 124514, with a year's production of 27,761.7 lbs. milk, testing 4.34, and 1205.09 lbs. butter fat.

The native home of the Holstein-Friesian cattle is Holland. Their introduction into America dates back to the 17th century. The Holstein-Friesian Association of America was organized to promote the breed in 1885. F. L. Houghton, Brattleboro, Vt., is the present secretary. M. H. Gardner, Delavan, Wis., supervises the registration of Advanced Registry animals.

SCALE OF POINTS FOR HOLSTEIN COWS

Head.—Decidedly feminine in appearance; fine in contour.................................. 2
    Discredit v. s. ⅛-s. ¼-m. ½-v. m. ¾-e. 1.
Forehead.—Broad between the eyes; dishing.. 2
    Discredit v. s. ⅛-s. ¼-m. ½-v. m. ¾-e. 1.
Face.—Of medium length; clean and trim, especially under the eyes, showing facial veins; the bridge of the nose straight.. 2
    Discredit s. ⅛-m. ¼-e. ½.
Muzzle.—Broad with strong lips... 1
    Discredit m. ⅛-e. ¼.
Ears.—Of medium size; of fine texture; the hair plentiful and soft; the secretions oily and abundant... 1
    Discredit m. ⅛-e. ¼.
Eyes.—Large; full; mild; bright.. 2
    Discredit s. ⅛-m. ¼-e. ½.
Horns.—Small; tapering finely towards the tips; set moderately narrow at base; oval; inclining forward; well bent inward; of fine texture; in appearance waxy... 1
    Discredit m. ⅛-e. ¼.
Neck.—Long; fine and clean at juncture with the head; free from dewlap; evenly and smoothly joined to shoulders... 4
    Discredit v. s. ⅛-s. ¼-m. ½-v. m. ¾-e. 1.
Shoulders.—Slightly lower than hips; fine and even over tops; moderately broad and full at sides... 3
    Discredit v. s. ⅛-s. ¼-m. ½-v. m. ¾-e. 1.

---

*In old cows some allowance should be shown in view of the fact that bearing calves from year to year tends to weaken them in this matter.

Chest.—Of moderate depth and lowness; smooth and moderately full in the brisket, full in the foreflanks (or through at the heart).................................... 6
    Discredit v. s. ¼-s. ½-m. 1-v. m. 1½-e. 2.

Crops.—Moderately full... 2
    Discredit v. s. ¼-s. ½-m. 1-v. m. 1½-e. 2.

Chine.—Straight; strong; broadly developed, with open vertebrae..................... 6
    Discredit v. s. ⅛-s. ¼-m. ½-v. m. ¾-e. 1.

Barrel.—Long; of wedge shape; well rounded; with a large abdomen, trimly held up, (in judging the last item age must be considered)*............................. 7
    Discredit v. s. ⅛-s. ¼-m. ½-v. m. ¾-e. 1.

Loin and Hips.—Broad; level or nearly level between hook bones; level and strong laterally; spreading from the chine broadly and nearly level; hook bones fairly prominent... 6
    Discredit v. s. ⅛-s. ¼-m. ½-v. m. ¾-e. 1.

Rump.—Long; high; broad with roomy pelvis; nearly level laterally; comparatively full above the thurl; carried out straight to dropping of tail........ 6
    Discredit v. s. ⅛-s. ¼-m. ½-v. m. ¾-e. 1.

Thurl.—High; broad.. 3
    Discredit v. s. ¼-s. ½-m. 1-v. m. 1½-e. 2.

Quarters.—Deep; straight behind; twist filled with development of udder; wide and moderately full at the sides... 4
    Discredit v. s. ⅛-s. ¼-m. ½-v. m. ¾-e. 1.

Flanks.—Deep; comparatively full... 2
    Discredit v. s. ⅛-s. ¼-m. ½-v. m. ¾-e. 1.

Legs.—Comparatively short; clean and nearly straight; wide apart; firmly and squarely set under the body; feet of medium size, round, solid and deep........ 4
    Discredit v. s. ⅛-s. ¼-m. ½-v. m. ¾-e. 1.

Tail.—Large at base, the setting well back; tapering finely to switch; the end of the bone reaching the hocks or below; the switch full..................................... 2
    Discredit s. ⅛-m. ¼-e. ½.

Hair and Handling.—Hair healthful in appearance; fine, soft and furry; the skin of medium thickness and loose; mellow under the hand; the secretions oily, abundant and of a rich brown or yellow color.. 8
    Discredit v. s. ¼-s. ½-m. 1-v. m. 1½-e. 2.

Mammary Veins.—Very large; very crooked (age must be taken into consideration on judging of size and crookedness); entering very large or numerous orifices; double extensions; with special developments such as branches, connections, etc.. 10
    Discredit, v. s. ¼-s, ½-m, 1-v, m. 1½-e. 2.

Udder and Teats.—Very capacious; very flexible; quarters even; nearly filling the space in the rear below the twist; extending well forward in front; broad and well held up.. 12
    Discredit v. s. ¼-s. ½-m. 1-v. m. 1½-e. 2.

Teats.—Well formed; wide apart, plumb and of convenient size....................... 2
    Discredit s. ⅛-m. ¼-e. ½.

Escutcheon.—Largest; finest... 2
    Discredit s. ⅛-m. ¼-e. ½.

                                                                                                             100

General Vigor.—For deficiency inspectors may discredit from the total received, not to exceed eight points.
    Discredit v. s. 1-s. 2-m. 3-v. m. 5-e. 8.

General Symmetry and Fineness.—For deficiency inspectors may discredit from the total received, not to exceed eight points.
    Discredit v. s. 1-s. 2-m. 3-v. m. 5-e. 8.

General Style and Bearing.—For deficiency inspectors may discredit from the total received, not to exceed eight points.
    Discredit v. s. 1-s. 2-m. 3-v. m. 5-e. 8.

Credits for Excess of Requirement in Production.—A cow shall be credited one point in excess of what she is otherwise entitled to, for each and every 10 per cent that her butter-fat record exceeds the minimum requirements for her age.

In scaling for the Advanced Register, defects caused solely by age, or by accident, or by disease not hereditary, shall not be considered. But in scaling for the show-ring such defects shall be considered and duly discredited.

A cow that, in the judgment of the inspector, will not reach at full age in milking condition and ordinary flesh 1,000 pounds live weight, and scale at least seventy-five points, shall be disqualified for entry, with description, in the Advanced Register.

FIG. 24.—STARTED ON ANOTHER OF THE CHANNEL ISLANDS

Jersey cow Croatia of St. Lambert, No. 218,608. Her year's production is 14,499.2 lbs. milk, testing 4.89 and 709.22 lbs. butter fat, the highest production of any cow of her breed in Wisconsin.

Sophie 19th of Hood Farm, No. 189748, with a production of 17,557.8 lbs. milk, testing 5.69 per cent and 999.14 lbs. butter fat, holds the world's record for this breed.

The native home of Jersey cattle is on the Island of Jersey in the English Channel. Jersey cattle were imported into America as early as 1818. The American Jersey Cattle Club, was organized in 1868. R. M. Gow, 324 W. 23rd St., New York City is the present secretary.

SCALE OF POINTS FOR JERSEY COWS

Head		7
A. Medium size, lean; face dished; broad between eyes and narrow between horns	4	
B. Eyes full and placid; horns small to medium, incurving; muzzle broad, with muscular lips; strong under jaw	3	
Neck.—Thin, rather long, with clean throat; thin at withers		5
Body		33
A. Lung capacity, as indicated by depth and breadth through body, just back of fore legs	5	
B. Wedge shape, with deep, large paunch; legs proportionate to size, and of fine quality	10	
C. Back straight to hip-bones	2	
D. Rump long to tail-setting and level from hip-bones to rump-bones	8	
E. Hip-bones high and wide apart; loins broad, strong	5	
F. Thighs flat and well cut out	3	
Tail.—Thin, long, with good switch, not coarse at setting on		2
Udder		28
A. Large size and not fleshy	6	
B. Broad, level or spherical, not deeply cut between teats	4	
C. Fore udder full and well rounded, running well forward of front teats	10	
D. Rear udder well rounded, and well out and up behind	8	
Teats.—Of good and uniform length and size, regularly and squarely placed		8
Milk Veins.—Large, tortuous and elastic		4
Size.—Mature cows, 800 to 1,000 pounds		3
General Appearance		10
A. A symmetrical balancing of all the parts, and a proportion of parts to each other, depending on size of animal, with the general appearance of a high-class animal, with capacity for food and productiveness at pail	10	
		100

EXPERIMENT STATION STAFF

The President of the University

H. L. Russel, Dean and Director
F. B. Morrison, Asst. Dir. Expt. Station
K. L. Hatch, Asst. Dir. Agr. Extension Service
C. W. Vaughn, Executive Secretary

W. A. Henry, *Emeritus* Agriculture
S. M. Babcock, *Emeritus* Agr. Chemistry
A. S. Alexander, Veterinary Science; in charge of station Enrollment
F. A. Aust, Horticulture
B. A. Beach, Veterinary Science
G. H. Benkendorf, Dairy Husbandry
Cora E. Binzel, Home Economics
D. S. Bullock, Animal Husbandry
L. J. Cole, In charge of Experimental Breeding
Amy L. Daniels, Home Economics
E. J. Delwiche, Agronomy (Ashland)
E. H. Farrington, In charge of Dairy Husbandry
E. B. Fred, Agr. Bacteriology
W. D. Frost, Agr. Bacteriology
J. G. Fuller, Animal Husbandry
C. J. Galpin, Country Life Work
W. J. Geib, Soils
F. B. Graber, Agronomy
F. B. Hadley, In charge of Veterinary Science
J. G. Halpin, In charge of Poultry Husbandry
E. B. Hart, In charge of Agr. Chemistry
E. G. Hastings, In charge of Agr. Bacteriology
K. L. Hatch, In charge of Agr. Education
B. H. Hibbard, Agr. Economics
A. W. Hopkins, Editor; In charge of Agr. Journalism
G. C. Humphrey, In charge of Animal Husbandry
J. A. James, Agr. Education
A. G. Johnson, Plant Pathology
J. Johnson, Horticulture
E. R. Jones, Soils
L. R. Jones, In charge of Plant Pathology
Elizabeth B. Kelley, Home Economics
G. W. Keitt, Plant Pathology
F. Kleinheinz, Animal Husbandry
B. D. Leith, Agronomy (Highland)
C. D. Livingston, Agr. Engineering
Abby L. Marlatt, In charge of Home Economics
E. V. McCollum, Agr. Chemistry
J. G. Milward, Horticulture
J. G. Moore, In charge of Horticulture
R. A. Moore, In charge of Agronomy
F. B. Morrison, Animal Husbandry
F. L. Musbach, Soils (Marshfield)
A. C. Oosterhuis, Animal Husbandry
D. H. Otis, Farm Management
W. H. Peterson, Agr. Chemistry
J. L. Sammis, Dairy Husbandry
Celestine Schmit, Home Economics
H. Steenbock, Agr. Chemistry
H. W. Stewart, Soils
A. L. Stone, Agronomy; In charge of Seed Inspection
H. C. Taylor, In charge of Agr. Economics
J. L. Tormey, Animal Husbandry
W. E. Tottingham, Agr. Chemistry
E. Truog, Soils
R. E. Vaughan, Plant Pathology
H. L. Walster, Soils
W. W. Weir, Soils
F. M. White, In charge of Agr. Engineering, pro tem.
A. R. Whitson, In charge of Soils
H. F. Wilson, In charge of Economic Entomology
J. F. Wojta, Field Supervisor Extension Courses and Schools
W. H. Wright, Agr. Bacteriology

A. R. Albert, Soils
Freda Bachmann, Agr. Bacteriology
W. L. Bevan, Economic Entomology
J. A. Becker, Agr. Economics
T. L. Bewick, Agr. Extension
J. W. Brann, Horticulture
Florence M. Coerper, Plant Pathology
J. H. Coffman, Vet. Science
A. C. Dahlberg, Dairy Husbandry
H. A. Drescher, Agr. Chemistry
Chas. L. Fluke, Economic Entomology
H. Fulmer, Agr. Bacteriology
J. J. Garland, Agronomy
L. M. Gentner, Economic Entomology
Dwight Getchell, Feed and Fertilizer Inspection
E. J. Graul, Soils
C. I. Griffith, Agr. Engineering
R. V. Gunn, Agr. Economics
L. P. Hanson, Soils
R. T. Harris, Dairy Tests
J. B. Hayes, Poultry Husbandry
C. S. Hean, Agr. Library
J. R. Hepler, Horticulture
H. Ibsen, Experimental Breeding
O. N. Johnson, Poultry Husbandry
F. R. Jones, Agr. Eng.
H. M. Jones, Agronomy (Ashland)
Myrtle Jones, Home Economics
O. A. Juve, Agr. Economics
F. J. Kelley, Experimental Breeding
A. H. Kuhlman, Animal Husbandry
F. D. Lewis, Asst. to the Dean
H. Lunz, Agronomy
O. G. Malde, Cranberry Investigations (Grand Rapids)
W. E. Markey, Animal Husbandry
E. R. McIntyre, Agr. Journalism
G. B. Mortimer, Agronomy
Reid F. Murray, Agr. Extension
F. E. Mussehl, Poultry
V. E. Nelson, Agr. Chemistry
W. Pitz, Agr. Chemistry
G. Potter, Horticulture
R. H. Roberts, Horticulture
H. H. Roehm, Agr. Bacteriology
E. C. Sauve, Agr. Engineering
A. A. Schaal, Agr. Chemistry
Nina Simmonds, Agr. Chemistry
J. E. Simmons, Agr. Bacteriology
Elizabeth Staley, Agr. Journalism
E. Steigleder, Experimental Breeding
W. H. Strowd, Feed and Fertilizer Inspection
W. A. Sumner, Agr. Journalism
H. W. Ullsperger, Soils (Sturgeon Bay)
H. O. Watrud, Agr. Economics
G. D. Williams, Feed and Fertilizer Inspector
C. M. Woodworth, Experimental Breeding
A. H. Wright, Agronomy

AGRICULTURAL REPRESENTATIVES

E. L. Luther, State Supervisor
Geo. M. Briggs, Burnett county
A. H. Cole, Lincoln county
J. M. Coyner, Portage county
R. L. Cuff, Barron county
Oscar Gunderson, Vilas county
G. M. Householder, Rusk county
G. R. Ingalls, Eau Claire county
W. D. Juday, Oneida county
J. S. Klinka, Polk county
R. A. Kolb, Taylor county
L. L. Oldham, Walworth county
C. B. Post, Ashland county
Griffith Richards, Price county
John Swenehart, Forest county
F. G. Swoboda, Langlade county
John Waiz, Douglas county
C. P. West, Sawyer county